WHAT IS STEAM?

WHAT ARE STEAM JOBS?

BY THERESA EMMINIZER

Please visit our website, www.garethstevens.com. For a free color catalog of all our high-quality books, call toll free 1-800-542-2595 or fax 1-877-542-2596.

Cataloging-in-Publication Data
Names: Emminizer, Theresa.
Title: What are STEAM jobs? / Theresa Emminizer.
Description: New York : Gareth Stevens Publishing, 2024. | Series: What is STEAM? | Includes glossary and index.
Identifiers: ISBN 9781538285596 (pbk.) | ISBN 9781538285602 (library bound) | ISBN 9781538285619 (ebook)
Subjects: LCSH: Science–Vocational guidance–Juvenile literature. | Technology–Vocational guidance–Juvenile literature. | Engineering–Vocational guidance–Juvenile literature. | Art–Vocational guidance-Juvenile literature. | Mathematics–Vocational guidance–Juvenile literature.
Classification: LCC Q147.E46 2023 | DDC 502.3–dc2

Published in 2024 by
Gareth Stevens Publishing
2544 Clinton Street
Buffalo, NY 14224

Copyright © 2024 Gareth Stevens Publishing

Designer: Leslie Taylor
Editor: Theresa Emminizer

Photo credits: Series Art (background art) N.Savranska/Shutterstock.com; Cover Dragon Images/Shutterstock.com; p. 5 Basileus/Shutterstock.com; p. 7 Frecca/Shutterstock.com; p. 9 Ground Picture/Shutterstock.com; p. 11 Dmytro Zinkevych/Shutterstock.com; p. 13 Artur Didyk/Shutterstock.com; p. 15 Andrey_Popov/Shutterstock.com; p. 17 wavebreakmedia/Shutterstock.com; p. 19 Yuricazac/Shutterstock.com; p. 21 Djavan Rodriguez/Shutterstock.com.

All rights reserved. No part of this book may be reproduced in any form without permission in writing from the publisher, except by a reviewer.

Printed in the United States of America

Some of the images in this book illustrate individuals who are models. The depictions do not imply actual situations or events.

CPSIA compliance information: Batch #CSGS24: For further information contact Gareth Stevens at 1-800-542-2595.

CONTENTS

What Will You Do?................ 4
Science Jobs..................... 6
Technology Jobs................. 8
Engineering Jobs................ 10
Art Jobs........................ 12
Math Jobs....................... 14
STEAM Skills.................... 16
Do What You Love................ 20
Glossary........................ 22
For More Information............ 23
Index........................... 24

Boldface words appear in the glossary.

What Will You Do?

What do you want to be when you grow up? That's a big question to answer! Thinking about what you like is a good place to start. Are you into **STEAM**? In our fast-changing world, there's a growing need for STEAM workers!

Science Jobs

A scientist's work is all about understanding the world around us. Scientists learn by doing **research** and **experiments**. There are many scientific fields, or areas of study. For example, zoologists study animals. They often work in zoos and wildlife parks.

Technology Jobs

Technology work is about inventing, or making, useful tools and finding answers to problems. Science, math, and engineering are often a big part of technology jobs. If you love spending time on your computer, you might like a job in technology!

CONTENTS

What Will You Do? 4
Science Jobs 6
Technology Jobs 8
Engineering Jobs 10
Art Jobs 12
Math Jobs 14
STEAM Skills 16
Do What You Love 20
Glossary 22
For More Information 23
Index 24

Boldface words appear in the glossary.

What Will You Do?

What do you want to be when you grow up? That's a big question to answer! Thinking about what you like is a good place to start. Are you into **STEAM**? In our fast-changing world, there's a growing need for STEAM workers!

Science Jobs

A scientist's work is all about understanding the world around us. Scientists learn by doing **research** and **experiments**. There are many scientific fields, or areas of study. For example, zoologists study animals. They often work in zoos and wildlife parks.

Technology Jobs

Technology work is about inventing, or making, useful tools and finding answers to problems. Science, math, and engineering are often a big part of technology jobs. If you love spending time on your computer, you might like a job in technology!

Engineering Jobs

Engineers use science and math to build better objects, or things. Engineers plan and build machines, systems, and **structures**, from cars to roads to bridges! If you like building and **designing**, you might want to be an engineer.

Art Jobs

Artists use their **imaginations** to find new ways of sharing their thoughts, feelings, and ideas with others. Some do this by drawing and painting, others by dancing or making music! Art is about **creating** something that shares your feelings with others.

Math Jobs

Math is the study of numbers. Mathematicians look for **patterns** in numbers and come up with new ways of answering problems and doing things. There are many different math jobs. For example, finance work is about studying money and how it's spent.

STEAM Skills

Whether it's work in science, technology, engineering, art, or math, all STEAM jobs are about exploring. That means searching in order to find out new things! Across the many fields of study, STEAM workers share a set of skills.

People who practice STEAM must be **curious**. They must ask questions. They must experiment and think outside the box to come up with answers. Above all, they must be ready to make mistakes and start over again.

Do What You Love

What interests you most? Maybe you light up when you learn about science! Maybe you wonder about how things work or daydream about finding better ways to get things done. Maybe you love to take pictures. Could a STEAM job be right for you?

GLOSSARY

create: To make something.

curious: Wanting to know or learn something.

design: To create the pattern or shape of something.

experiment: To carry out scientific tests or actions to learn about something.

imagination: A place in the mind where you picture things or come up with ideas.

pattern: The way something happens over and over again.

research: Studying to learn more about something.

STEAM: Stands for Science, Technology, Engineering, Art, and Math.

structure: Something built.

FOR MORE INFORMATION

BOOKS

Brundle, Joanna. *My Job in Math.* New York, NY: PowerKids Press, 2022.

Brundle, Joanna. *My Job in Science.* New York, NY: PowerKids Press 2022.

WEBSITES

NASA Science
spaceplace.nasa.gov/science/en/
Learn how to think scientifically by asking questions and making observations.

Science Buddies
www.sciencebuddies.org/science-fair-projects/project-ideas/first-grade
Have fun with STEAM! Try out these experiments by Science Buddies.

Publisher's note to educators and parents: Our editors have carefully reviewed these websites to ensure that they are suitable for students. Many websites change frequently, however, and we cannot guarantee that a site's future contents will continue to meet our high standards of quality and educational value. Be advised that students should be closely supervised whenever they access the internet.

INDEX

animal, 6
art, 12, 16
building, 10
computer, 8
dancing, 12
drawing, 12
engineering, 8, 10, 16
experiment, 6, 18
finance, 14
machine, 10
math, 8, 10, 14, 16
painting, 12
research, 6
science, 6, 8, 10, 16, 20
skill, 16
structure, 10
technology, 8, 16
zoo, 6